MATHAIM
PROBLEM-SOLVING
WORKBOOK
GRADES 4-6 (LOGIC)
VOLUME 1

This publication is designed to provide accurate and authoritative information in regard to the subject matter covered. It is sold with the understanding that the publisher is not engaged in rendering legal, accounting, or other professional service. If legal advice or other expert assistance is required, the services of a competent professional should be sought.

Copyright 2012 Math AIM.

All rights reserved. The text of this publication, or any part thereof, may not be reproduced in any manner whatsoever without written permission from the publisher.

Printed in the United States of America.

ISBN 13: 978-1479280544

Formula/Tips:

1. A box of marbles weighs 70 grams when full and 50 grams when half full. How many grams does the box weigh when empty?

2. A container of beans weighs 60 grams when full and 50 grams when half full. How many grams does the container weigh when empty?

3. A lunch box weighs 90 grams when full and 50 grams when half full. How many grams does the lunch box weigh when empty?

4. A basket of fruits weighs 140 grams when full and 90 grams when half full. How many grams does the basket weigh when empty?

5. A basket of laundry weighs 540 grams when full and 310 grams when half full. How many grams does the basket weigh when empty

Formula/Tips:

6. The volunteering club has 14 members. Two members go to the elementary school to help every school day. What is the greatest number of school days that can pass without repeating the same pair of students?

7. The P.E. club has 10 members. Two members take attendance every school day. What is the greatest number of school days that can pass without repeating the same pair of students?

8. The cafeteria has 18 members. 3 people go to the store to buy the food every day. What is the greatest number of days that can pass without repeating the same group of people?

9. The track team has 32 members. 4 people run a race every day. What is the greatest number of days that can pass without repeating the same group of people?

10. The camp has 17 members. 3 people go out every day to find food. What is the greatest number of days that can pass without repeating the same group of people

Formula/Tips:

11. A B   If A and B are different digits, what is the 3 digit product?
x B A
    □ 2
  □□
 □ □□

12. F G   If F and G are different digits, what is the 3 digit product?
X G F
    □ 5
  □□
 □ □□

13. X Y   If X and Y are different digits, what is the three digit product?
X Y X
    □ 6
  □□
 □ □□

14. P Q   If P and Q are different digits, what is the three digit product?

X   Q P

    ☐ ☐ 3
  ☐ ☐ ☐
  ☐ ☐ ☐ ☐

15. C D   If C and D are different digits, what is the three digit product?

X   D C

    ☐ 4
  ☐ ☐
  ☐ ☐ ☐

Formula/Tips:

16. The product of 506 and 804 is divided by the sum of 34 and 66. What is the remainder?

17. The product of 607 and 404 is divided by the sum of 42 and 58. What is the remainder?

18. The product of 506 and 908 is divided by the sum of 76 and 24. What is the remainder?

19. The product of 309 and 811 is divided by the sum of 84 and 16. What is the remainder?

20. The product of 4,006 and 7,010 is divided by the sum of 235 and 765. What is the remainder?

Formula/Tips:

21. A fenced rectangular garden is 5 m wide and 12 m long. When one side is moved outward and two other sides are increased in length, the area increases by 60 sq m. What is the fewest number of meters of additional fencing needed to form the larger rectangular garden?

22. A fenced rectangular garden is 4 m wide and 8 m long. When one side is moved outward and two other sides are increased in length, the area increases by 40 sq m. What is the fewest number of meters of additional fencing needed to form the larger rectangular garden?

23. A fenced rectangular garden is 9 m wide and 15 m long. When one side is moved outward and two other sides are increased in length, the area increases by 45 sq m. What is the fewest number of meters of additional fencing needed to form the larger rectangular garden?

24. A fenced rectangular garden is 12 m wide and 18 m long. When one side is moved outward and two other sides are increased in length, the area increases by 36 sq m. What is the fewest number of meters of additional fencing needed to form the larger rectangular garden?

25. A fenced rectangular garden is 8 m wide and 16 m long. When one side is moved outward and two other sides are increased in length, the area increases by 32 sq m. What is the fewest number of meters of additional fencing needed to form the larger rectangular garden?

Formula/Tips:

26. What is the value of 23 x 14 + 45 x 14 + 32 x 14?

27. What is the value of 34 x 19 + 43 x 19 + 23 x 19?

28. What is the value of 16 x 26 + 32 x 26 + 52 x 26?

29. What is the value of 43 x 37 + 25 x 37 + 32 x 37?

30. What is the value of 235 x 69 + 445 x 69 + 320 x 69?

Formula/Tips:

31. Billy began with a whole number. He divided the number by 2, subtracted 6 from the quotient, took the square root of the difference, added 14 to the square root, and took the square root of the sum. His final result was 4. What was Billy's original number?

32. Tom Canty began with a whole number. He divided the number by 2, subtracted 10 from the quotient, took the square root of the difference, added 34 to the square root, and took the square root of the sum. His final result was 8. What was Tom's original number?

33. Sara began with a whole number. She divided the number by 3, subtracted 4 from the quotient, took the square root of the difference, added 36 to the square root, and took the square root of the sum. Her final result was 10. What was Sara's original number?

34. Ria began with a whole number. She divided the number by 2, subtracted 8 from the quotient, took the square root of the difference, added 15 to the square root, and took the square root of the sum. Her final result was 5. What was Sara's original number?

35. Divya began with a whole number. She divided the number by 9, subtracted 9 from the quotient, took the square root of the difference, added 40 to the square root, and took the square root of the sum. Her final result was 7. What was Divya's original number?

Formula/Tips:

36. What is the sum of all the prime numbers between 10 and 30?

37. What is the sum of all the prime numbers between 30 and 50?

38. What is the sum of all the prime numbers between 60 and 70?

39. What is the sum of all the prime numbers between 80 and 90?

40. What is the sum of all the prime numbers between 70 and 100?

Formula/Tips:

41. The average of five numbers is 15. Four of the numbers are 13, 17, 15, and 12. Find the other number.

42. The average of five numbers is 43. Four of the numbers are 32, 23, 54, and 49. Find the other number.

43. The average of five numbers is 64. Four of the numbers are 94, 82, 57, and 23. Find the other number.

44. The average of five numbers is 93. Three of the numbers are 111, 118, and 108. The fifth number is 18 more than the fourth number. Find the fourth and fifth number.

45. The average of five numbers is 117. Three of the numbers are 143, 156, and 162. The fifth number is 38 more than the fourth number. Find the fourth and fifth number

Formula/Tips:

46. The letters in **Forty** and **One** are cycled <u>separately</u> as shown and placed in a numbered vertical list. After line 1, the <u>next</u> line in which both **Forty** and **One** are spelled correctly is line *N*. Find *N*.

1. Forty One
2. yFort eOn
3. tyFor neO
....
N. Forty One

47. The letters in **Fifty** and **Nine** are cycled <u>separately</u> as shown and placed in a numbered vertical list. After line 1, the <u>next</u> line in which both **Fifty** and **Nine** are spelled correctly is line *X*. Find *X*.

1. Fifty Nine
2. yFift eNin
3. tyFif neNi
....
X. Fifty Nine

48. The letters in **Eighty** and **Two** are cycled <u>separately</u> as shown and are placed in a numbered vertical list. After line 1, the <u>next</u> line in which both **Eighty** and **Two** are spelled correctly is line *Q*. Find *Q*.

1. Eighty Two
2. yEight oTw
3. tyEigh woT
....
Q. Eight Two

49. P = **31**

50. Z = **13**

Formula/Tips:

51. A telephone call costs 28¢ for the first 2 minutes and 5¢ for each additional minute. If Joe Curry pays 58¢ for a call, for how many minutes did the call last?

52. A telephone call costs 30¢ for the first three minutes and 6¢ for each additional minute. If Little Timmy pays 90¢ for a call, for how many minutes did the call last?

53. A telephone call costs 35¢ for the first 4 minutes and 9¢ for each additional minute. If Rohit pays 71¢ for a call, for how many minutes did the call last?

54. A telephone call costs 20¢ for the first 4 minutes and 6¢ for each additional minute. If Abhishek pays 116¢ for a call, for how many minutes did the call last?

55. A telephone call costs 32¢ for the first 4 minutes and 4¢ for each additional minute. If Madam John pays 120¢ for a call, for how many minutes did the call last?

Formula/Tips:

56. The product of two whole numbers is 44. The average of the two numbers is 12. Find the greater of the numbers.

57. The product of two whole numbers is 45. The average of the two numbers is 7. Find the least of the numbers.

58. The product of two whole numbers is 81. The average of the two numbers is 15. Find the difference between the numbers.

59. The product of two whole numbers is 100. The average of the two numbers is 10. Find the difference between the numbers.

60. The product of two whole numbers is 231. The difference between the two numbers is 26. Find the average of the numbers.

Formula/Tips:

61. *OH* and *HO* represent two 2-digit numbers. If *OH – HO* = 18, what is the value of the expression $O - H$?

62. *GO* and *OG* represent two 2-digit numbers. If *GO – OG* = 27, what is the value of the expression $G - O$?

63. *JOB* and *BOJ* represent two 3-digit numbers. If *JOB – BOJ* = *B97*, what is the value of the expression $OB - BO$?

64. *LOL* and *OLL* represent two 3-digit numbers. If *LOL +OLL* = *LL02*, what is the value of the expression $OL - OL$?

65. *OMG* and *GMO* represent two 3-digit numbers. If *OMG + GMO* = *7GM*, what is the *average* of *OMG* and *GMO*?

Formula/Tips:

66. Find the sum of the counting numbers from 1 to 45

67. Find the sum of the counting numbers from 1 to 125

68. Find the sum of the counting numbers from 1 to 458.

69. Find the sum of the counting numbers from 1 to 1245.

70. Find the sum of the counting numbers from 1 to 56.

Formula/Tips:

71. A book has number 1 to 600, how many times does the digit 3 appears in the page numbers?

72. A book has number 1 to 800, how many times does the digit 5 appears in the page numbers?

73. A book has number 1 to 900, how many times does the digit 0 appears in the page numbers?

74. A book has number 1 to 1000, how many times does the digit 1 appears in the page numbers?

75. A book has number 1 to 950, how many times does the digit 7 appears in the page numbers?

Formula/Tips:

76. A box of candy can be divided in equal shares among 2, 4, 5 or 7 children with no candy left over. What is the least number of candies the box could have?

77. A box of candy can be divided in equal shares among 3,5,or 7 children with no candy left over. What is the least number of candies the box could have?

78. A box of candy can be divided in equal shares among 2,3,6 or 8 children with no candy left over. What is the least number of candies the box could have?

79. A box of candy can be divided in equal shares among 4,5,or 6 children with no candy left over. What is the least number of candies the box could have?

80. A box of candy can be divided in equal shares among 3,4,5,6 or 9 children with no candy left over. What is the least number of candies the box could have?

Formula/Tips:

---

81. If today is Monday. What day of the week will it be 74 days from now?

82. If today is Sunday. What day of the week will it be 175 days from now?

83. If today is Friday. What day of the week will it be 564 days from now?

84. If today is Wednesday. What day of the week will it be 900 days from now?

85. If today is Thursday. What day of the week will it be 865 days from now?

Formula/Tips:

86. The four digit number 5BB7 is divisible by 3. What digits does B represent?

87. The four digit number 65X8 is divisible by 4. What digits does X represent?

88. The four digit number 9BB7 is divisible by 9. What digits B represent?

89. The four digit number 8BB4 is divisible by 6. What digits B represent?

90. The four digit number 6BB1 is divisible by 3. What digits B represent?

Formula/Tips:

91. In a shop, pencils and pens have different prices. 3 Pencils and 5 pens cost 84¢. 5 pencils and 3 pens cost 76¢. How much does one pen and one pencil cost?

92. In a shop, pencils and pens have different prices. 4 Pencils and 8 pens cost 168¢. 8 pencils and 4 pens cost 156¢. How much does one pen and one pencil cost?

93. In a shop, pencils and pens have different prices. 7 Pencils and 5 pens cost 104¢. 5 pencils and 7 pens cost 112¢. How much does one pen and one pencil cost?

94. In a shop, pencils and pens have different prices. 5 Pencils and 6 pens cost 145¢. 6 pencils and 5 pens cost 141¢. How much does one pen and one pencil cost?

95. In a shop, pencils and pens have different prices. 10 Pencils and 8 pens cost 310¢. 8 pencils and 10 pens cost 320¢. How much does one pen and one pencil cost?

Formula/Tips:

96. In three games Pranathi scored 140, 130 and 135. How much she has to score in order to get an average of 137?

97. In three games Sumedha scored 230, 223 and 245. How much she has to score in order to get an average of 235?

98. In four games Divya scored 123, 167, 150 and 145. How much she has to score in order to get an average of 145?

99. In four games Samrit scored 156, 139,134, and 148. How much he has to score in order to get an average of 140?

100. In five games Harin scored 142,154, 128 and 135. How much does she have to score in order to get an average of 143?

Practice Test:

1. A box of marbles weighs 175 grams when full and 100 grams when half full. How many grams does the box weigh when empty?

2. The volunteering club has 13 members. 4 members go to the elementary school to help every school day. What is the greatest number of school days that can pass without repeating the same pair of students?

3. A B   If A and B are different digits, what is the 4 digit product?

   x B A

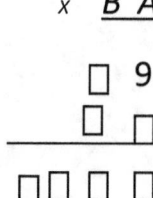

4. The product of 201 and 503 is divided by the sum of 734 and 266. What is the remainder?

5. A fenced rectangular garden is 5 m wide and 11 m long. When one side is moved outward and two other sides are increased in length, the area increases by 88 sq m. What is the fewest number of meters of additional fencing needed to form the larger rectangular garden?

6. What is the value of 24 x 45 + 32 x 45 + 44 x 45?

7. Ayush began with a whole number. He divided the number by 3, subtracted 5 from the quotient, took the square root of the difference, added 7 to the square root, and took the square root of the sum. His final result was 4. What was Tom's original number?

8. What is the sum of all the prime numbers between 65 and 85?

9. The average of five numbers is 200. Three of the numbers are 210, 170, and 185. The fifth number is 20 more than the fourth number. Find the fourth and fifth number.

10. A telephone call costs 25¢ for the first 6 minutes and 8¢ for each additional minute. If Rohit pays 230¢ for a call, for how many minutes did the call last?

11. The product of two whole numbers is 96. The average of the two numbers is 10. Find the difference between the numbers.

12. *JOB* and *BOJ* represent two 3-digit numbers. If *JOB – BOJ = B97,* what is the value of the expression *OB – BO?*

13. Find the sum of the counting numbers from 1 to 300.

14. A book has number 1 to 1000. How many times does the digit 2 appears in the page numbers?

15. A box of candy can be divided in equal shares among 2, 5, 6 or 8 children with no candy left over. What is the least number of candies the box could have?

16. If today is Sunday, what day of the week will it be 120 days from now?

17. The four digit number 8BB2 is divisible by 9. What digits does B represent?

18. In a shop, pencils and pens have different prices. 8 pencils and 4 pens cost 156¢. 4 Pencils and 8 pens cost 168¢. How much does one pen and one pencil cost?

19. In three games Sumedha scored 380, 360 and 395. How much does she have to score in order to get an average of 375?

20. The letters in **Ninety** and **Nine** are cycled <u>separately</u> as shown and placed in a numbered vertical list. After line 1, the <u>next</u> line in which both **Ninety** and **Nine** are spelled correctly is line N. Find N.

1. Ninety Nine
2. yNinet eNin
3. tyNine neNi
....
P. Ninety Nine

ANSWER KEY

1. 30	2. 40	3. 10	4. 40	5. 80
6. 91	7. 45	8. 816	9. 35960	10. 680
11. 252	12. 765	13. 736 or 976	14. 403	15. 574
16. 24	17. 28	18. 48	19. 99	20. 60
21. 10	22. 10	23. 6	24. 4	25. 4
26. 1,400	27. 1,900	28. 2,600	29. 3,700	30. 69,000
31. 20	32. 1820	33. 15000	34. 216	35. 810
36. 112	37. 199	38. 128	39. 172	40. 583
41. 18	42. 57	43. 64	44. 55,73	45. 43,81
46. 16	47. 21	48. 7	49. 31	50. 13
51. 8min.	52. 13min.	53. 8min.	54. 20	55. 26
56. 22	57. 5	58. 24	59. 0	60. 20
61. 2	62. 3	63. -18	64. 0	65. 363
66. 1035	67. 7875	68. 105,111	69. 775,635	70. 1596
71. 220	72. 260	73. 180	74. 301	75. 285
76. 140	77. 105	78. 24	79. 60	80. 180
81. Friday	82. Sunday	83. Tuesday	84. Sunday	85. Monday
86. 0,3,6,9	87. 2,0,4,8	88. 1	89. 0,3,6,9	90. 1,4,7
91. 20	92. 27	93. 19	94. 26	95. 35
96. 143	97. 242	98. 140	99. 123	100. 156

PRACTICE TEST

1. 25	2. 715	3. 1729	4. 103	5. 16
6. 45000	7. 162	8. 373	9. 1000	10. 10
11. 4	12. -18	13. 45150	14. 300	15. 120
16. Monday	17. 4	18. 27¢	19. 365	20. 13

MEET THE TEAM

From left to right: Pranathi, Sumedha, Raji Menon, Nick (the dog), Vivek, Samrit, Divya, Harin, Abhishek, Rohit

www.ingramcontent.com/pod-product-compliance
Lightning Source LLC
Chambersburg PA
CBHW081245180526
45171CB00005B/552